The Backseat Flyer

PLANE sense about flying as a passenger.

By

Nina Anderson

Illustrated by Richard Vail

Published by SAFE GOODS
East Canaan, CT 06024

The Backseat Flyer
by Nina Anderson
Illustrated by Richard Vail

Copyright© 1998 by Nina Anderson

ISBN 1884820-35-2-50995
Library of congress Catalog Card Number 98-90505
Printed in the United States of America

No part of The Backseat Flyer is intended as medical
advice. It is written solely for informational and educational
purposes. Please consult a health professional should the
need for one be indicated.

Published by SAFE GOODS
283 East Canaan, Rd
East Canaan, CT 06024
(860)-824-5301

DEDICATIONS

My part of this book is dedicated to my father, Nicholas Vuyosevich and my mother, Grace who because of their achievements in piloting airplanes, gave me the impetus to take up flying.
-Nina Anderson

I dedicate this book to my father, Richard Vail, Sr., a corporate pilot, who encouraged me to take my artistic talents seriously. My dad passed onto the cosmos at too early an age, but his love and interest in my work will never be forgotten.
-Richard Vail

FOREWORD

Nina Anderson and I have co-authored several books, and this one, "The Backseat Flyer" was definitely something I had to read! Many times I would call Nina and ask her questions about my upcoming flights around the country. For example, when I fly from coast to coast in the United States, I constantly have jet-lag. By the time I adapt to the time zone at my destination, I am on my way back home. I was happy that this book gave me some methods to reset my body clock quickly.

Another example of my problems with flying is my disgruntled attitude about flights being delayed or canceled. I find myself getting impatient and irritated at the airline. I really was enlightened when I read "Commandment number three: Don't Complain." I never really gave any thought to why those delays were happening and how my life would be on the line if they didn't fix what was broken before we took off. My biggest problem was fearing all those weird noises that occur while we are flying. I think the worst and the flight becomes a nightmare. Thank goodness, this book taught me that most of those are routine sounds and are nothing to worry about.

One day Nina called me and asked if I thought other frequent flyers went through the same emotions. From my conversations with the many passengers I flew with, I emphatically said YES! I also know many people who will not fly because of fear or because they have claustrophobia. Until I read this book, I never realized there were natural remedies that could alleviate the terror. I now carry a bottle of flower remedies with me in case my seatmate has a panic attack. Probably the most important message for me within these pages, was the information on The Great Gaping Mouth Syndrome. By not paying attention to the safety announcements, I was at risk to become one of those 149 passengers that may not have survived a loss of pressurization. I now am telling everyone to put on their oxygen mask if it pops out in front of them.... and ask questions later.

When my wife drives, I can honestly say that occasionally I am a *backseat driver*. Now, after reading this book I am happy to say I also am a *backseat flyer*. I eagerly look forward to my next flight, so I can apply the knowledge I found in this book.

Thank you Nina, for writing this book. It is a blessing for my serenity.

-Dr. Howard Peiper, O.D., Naturopath

INTRODUCTION

Did you ever complain about someone's driving when you were a passenger? Come on—admit it! We all have. And what did you have in common with the driver? You know how to drive and from your vantage point, you probably know how to drive BETTER than he does. This little fact qualifies you for the title of backseat driver.

Now, suppose you want to be a backseat flyer next time you take that little trip in an airplane. Does this mean you have to know how to fly the airplane to qualify? You bet it does—or how else can you know if the pilot is not living up to *your* standards. Since most of you haven't a clue how to fly an airplane, we aim to give you pointers so at least you can make criticisms where they are due!

A pilot must undergo initial hard work and tedious hours for little or no pay, and days upon days away from home. They also must confront the onslaught of job rejections because they haven't accumulated enough flight hours. As it takes lots of years and lots of money to get enough flying experience for someone to pay you to fly, you may wonder just why people choose a flying career. It's because the thrill gets in our blood and we will do anything for time in the air.

For many of us flying is genetic. Both my parents flew in a time when very, very, very few people ever left the ground. My mom flew solo and was on her way to being one of the few women in sport aviation, when her career was ended from a badly broken arm. My dad was a commercial aviator in the old smudge pot and rag wing days. Because of them, I was destined to be a pilot before I was even born.

TABLE OF CONTENTS

10

THE FLYING BUG.

The flying bug causes people to spend lots of money to defy gravity. Pilots and passengers put their trust in a machine that's built by somebody who is comfortably sitting in his living room. In my case, the flying bug bit me at home. My dad brainwashed me from the time I still had diapers on, giving me my first look out of an airplane window at 3 years old. That look was from the cockpit of a DC-3 and it must have planted the seed for me to get behind the controls! At that time the only women to fly airplanes were military ferry pilots and Amelia Erhardt. I was destined to forge new territory, as the 1970's saw a change in the prejudice of cockpit gender. Before we go forward, let's take a little trip back to my aviation roots.

There was a time when airplanes required scarves and goggles and runway lights were smudge pots. My dad knows, for he was a regular civilian pilot back in the thirties. The only aviation stories you hear about focus on the war hero or the record breaker. Let me tell you about the average person who wanted to fly back then.

Nicholas Vuyosevich (my dad) got his first ride during the "Gates Flying Circus Show" at Teterboro airport in New Jersey. It should have cost him five bucks, but my dad was a forlorn kid from the bowels of New York City (an area known as Hell's Kitchen) and only had two dollars to his name. At the end of the show one of the pilots felt sorry for dad and took him up. That was it. Nick got the flying bug.

It wasn't easy for a civilian to learn to fly in the 1920's. Airplanes were few and far between, and flight schools were even scarcer. To pursue his interest, my dad and a few other devotees started the Jersey City Flying Club. Sixty members made up the roster, but hardly any of them had a pilot's license. To realize their dream, they enticed a noteworthy aviator, Clarence Chamberlain to locate his flight school at their airport in Jersey City. Once accomplished, the flying club actually started flying. The only trouble was that Clarence feared the students would crash his biplanes and ordered his instructors never to let anyone fly solo. Therefore, my dad was logging lots of hours in a Kinner Swallow and Ox5 Swallow, but always with an instructor. Eventually he broke away and hired an independent instructor Chet Coons, and soloed on March 6, 1933. He received his private pilot's license shortly thereafter.

Nick Vuyosevich and his biplane.

What does a private pilot do in the 1930's— fly to Long Island for lunch? Mostly they tried to keep from crashing. Flying back then was not a precise art and landing up side down or sideways was a common occurrence. Fortunately, few pilots died in the crashes. My dad had greater sights and worked on his skills to obtain a commercial license that he received in 1939. He dreamed of being an airline pilot, and to that end flew commercial trips in aircraft such as a Staggerwing Beech. The great depression hit and private flying took a tumble. As jobs became scarce, airplanes were mothballed except for a few that my dad used in barnstorming air shows. When World War II started, many civilian pilots received their chance to fly bombers and fighters in the service. Not my dad. Although he was invited to fly with the famous wartime Flying Tigers, it was not meant to be. It seems he was a valuable employee of Walter Kidde Company, a defense contractor, and the government thought he needed to stay stateside *and grounded*. So ended my dad's flying career, but our house was never at a loss for numerous airplane pictures and old propellers that subliminally primed my brain towards aviation.

WHO ARE PILOTS.

I bet you don't know what we aviators (no gender intended) go through to sit up front and control your lives. Once we finish our training, we are constantly having our knowledge tested, whether it comes from airborne challenges or job interviews. We are faced with large amounts of rejection from possible employers who failed to hire us because we didn't have enough flight time, enough college credits, failed the ink blot test, or failed the shrink exam. In interviews for one major airline, the staff psychologist made us sit in a rocking chair. If the key to getting hired was to rock or not to rock, we never knew. Receiving constant rejection for employment makes pilots question the validity of setting goals and tests their self-esteem. But those of us that really believe and tough it out—make it.

At the start of my new venture into aviation, I had high expectations. I wore a T-shirt that said, *"When I grow up I want to be a pilot so I don't have to work."* That sounded good to me. Doing something I thought of as a hobby and getting paid big bucks for it, was my idea of the ultimate job. So I set a goal to become a pilot and thought about that day and night. I took numerous menial jobs to get money to pay for flying lessons. Three months later I flew solo. For those pilots reading this, I bet you can remember that day as if it was yesterday. Up to now you've been flying in an airplane that's not much bigger than a go-cart. The instructor in the seat next to you suddenly gets a sadistic side and suggests you fly ALONE.

You obligingly drop him off and taxi back to the runway. Nothing seems unusual until you get up in the air and realize that something is missing. The right seat is empty. There you are with nobody to bail you out if you screw up.

Talk about a growth experience. For the next five minutes you incessantly mutter all the things you have to do to get this machine back on the ground in one piece. You don't want to forget *anything*! Looking down at the airport while your hand is trying to hold onto the wheel, you feel you could win the trophy for sweaty palms, but what about the poor instructor. It's his neck on the line if you die, and he is totally helpless to do anything about it from the ground.

I survived my first solo, and when I landed I felt that I should be placed on the podium with a gold medal around my neck. This egotistic opinion of myself was short lived. A reality check soon found me going back to work at my menial job to make *more* money to keep flying. Two years later I earned my pilot's license, but something was wrong. I still had to keep paying to fly. What did I do wrong? Did my T- shirt lie?

I then discovered that the only people who make money at flying were those pilots who had a commercial license and an instrument rating. This meant that I had to take more flight instruction. More instruction meant spending more money. More money meant that I had to stay in my menial job. Once I received my advanced ratings, I discovered no one would hire me until I had a multi-engine rating. More money. More menial jobs. More ratings. Even though I now was qualified on paper, I lacked the flight time experience to land a job.

Needing money for food and shelter led me to become creative about building flight time for free. I was working as dispatcher for a charter airline and begged the owner to let me fly as an unpaid co-pilot. I'd arrive at the airport at 5 a.m. to fly a freight run. We would land back home at 8 a.m. I would work in the office until 5 when it would be back in the freighter for another run. The hours were killers, but they were preparing me for a lifetime career of getting up at "o-dark early" and flying until "o-dark late.

After a year and a half of "building time" I finally was hired as a captain to fly the freight runs. This meant that I was all by myself, *AGAIN*, fighting ice in the winter and thunderstorms in the summer. Our planes didn't have the luxury of radar and it makes you say your prayers when you blindly fly into a green cloud laced with lightning. By the way, do you know what the cockpit voice recorders reveal is the final words pilots say before they bite the dust? You guessed it—*oh shit!*

Well, I said lots of them, but luckily I didn't die. It did give me an undying respect for the power of weather. I eventually learned enough to become qualified to fly passengers, but my T-shirt was still lying! Sure, I was a pilot, but this was work and the pay was less than a fast food hamburger jockey gets. I decided that the guys who make the big bucks flying were the airline pilots. The problem with me being one of those, was that cockpits were named cockpit for a reason. Only Men Were Allowed. My window of opportunity was closed forever.

Then a wonderful thing happened. Equal rights hit aviation and women were in demand for pilot jobs. I finally was hired by an airline! I wore my uniform with pride and my ego became incorrigible until my first day on the job. Now, most of you have passed an airline pilot in the terminal. Do you treat this person with sanctity, awe and respect? That's what I thought this uniform commanded. After all I had spent lots of dollars on my training and employed large amounts of brain strain during flight school, so I deserved an ooooh or an aaaah! You passengers should think I'm cool. Well, it didn't take long for my bubble to burst. Do you know what kind of respect I got?

I repeat, **Do you know what respect I got?**

"Miss, where's the bathroom?" That's all. For all my years of slugging it out, spending money and studying hard, I get to tell passengers where the bathroom is. This made me realize that the only people who thought I was great, were the other pilots who also wore uniforms.

After a year of flying I realized I was still at the bottom of the pilot ladder. If any of you have reached what you thought was the pinnacle of your career, what did you find? Reality is not quite as good as what you though it would be. The airline I flew for was a "commuter" which meant we flew the smaller planes that people looked upon as death traps. Our salaries were minuscule compared to the jet pilots, *and* we had to load the baggage too. I wanted some respect, more money and I didn't want to load bags.

Companies that were big enough to need executive transportation, bought their own airplanes and established their own "airlines" complete with mechanics, flight attendants and pilots. In the 1970's, corporate flight departments had begun hiring women and I lucked out. I was among the first group of female pilots hired. We didn't wear ego boosting uniforms, but we had a fancy jet to fly, nice hotel rooms and lots more money. Other than when we flew international (in uniforms), we wore business suits and could be identified in the terminal using the Noah's Ark method. Look for pairs of pilots, wearing identical clothing and carrying a stainless coffee pot and a catering tray. Typical attire is blue blazers and tan trousers, or blue blazers and blue trousers, or blue blazers and gray trousers.

The Noah's Ark Pilot Identification Method

There are advantages to being a corporate pilot. One is that you get to fly to the Super Bowl. I remember one trip when we saw New England get hammered in New Orleans. You may not have 50 yard line tickets like we did, but at least you can say you flew a trip to the event. If you can imagine, there were 200 corporate jets parked on the ramp with 400 pilots (remember, we come in pairs) waiting the arrival of a thousand passengers. Picture a catering truck arriving and trying to get the right food to the right airplane. The parking ramp suddenly looked like the floor of the stock exchange. Caterers shouting out airplane ID numbers, pilots clambering over each other to get to their precious morsels, then making their way back to their ships to sit and wait for the herd of limousines to deliver their precious cargo. (Part of the challenge of corporate flying.)

Corporate flying was great. I was flying a jet and finally getting paid what I thought I was worth. My T-shirt finally was vindicated. I flew beautiful airplanes, received lots of respect and went on plenty of trips to warm, sunny beaches in the winter. For those years on the road, I learned about how to survive in the alien environment of a hotel room and an office at 35,000 feet.

You should have seen what I carried on the airplane with me. Not only did I carry hot clothes and cold clothes, a hair dryer and curling iron, night cream, day cream, make-up and my favorite crumpled up cotton nightie, but I also

brought my individual water purifier, my personal air purifier, my vitamins, my special wheat free cereal, and my own personal water bottle. Over the years, I identified many ways to avoid wrinkles, colds, gaining weight, and dehydration—all very real hazards of frequent flying!

I also identify with the unsuspecting passenger who gets white knuckles during *routine* flight procedures. In order to quell your worries and keep you healthy, we will take you through a flight. Sit back, and relax, but pay attention. Reading this book may be your answer to a life or death situation or at least reduce your fears about normal aircraft operating procedures.

The Six Commandments
of A Backseat Flyer.

✈

Commandment number one:
Worship the pilot.

✈

Commandment number two:
Listen.......to the safety announcements.

✈

Commandment number three:
Don't complain.

✈

Commandment number four:
**Know what all those whistles,
whirrs, bumps and clangs are
during a normal flight.**

✈

Commandment number five:
Obey the seat belt sign.

✈

Commandment number six:
**No booze, no caffeine, no carbonation,
your own water jug, a pack of gum
and a light box.**

Having come from experiences in both the back *and* the front of the airplane, I can empathize with the frequent traveler who has become blasé about flying and ignores the need to follow these commandments. Trust is a wonderful thing, especially trust in the pilot and the mechanic. This trust may not be enough to keep you from clenching the seat when the airplane instantaneously drops 100 feet—and then rises 200 feet. Or, when you know the ground is close at hand, but you can't see anything but fog out the window.

There are six commandments you <u>must</u> follow if you want to be a backseat flyer.

✈ *Commandment number one:*
Worship the pilot.

In an airplane, that person up front controls whether you live or die. You, the passenger, can be the top gun at the office or on the golf course. But, when you step on that airplane, you're nobody except a 170 pound entry on the weight and balance sheet. When things get tough, pilots don't even remember you're back there. They're too busy saving their own butt and yours just goes along for the ride.

Anyone who thinks that getting out of college means no more studying, should not take up flying. Pilots have to study *all the time*, go to retraining at least once a year, and are tested in simulators to make sure they remember what to do when the going gets tough.

Pilots have to do a lot of unglamorous things before they park their keisters in the airplane's front office. Pre-flighting the airplane is a must, to make sure the wings and wheels are still attached. The pilot checking the aircraft outside in the wind and rain is the junior pilot. The captain is comfortably sitting in the cockpit, where he is evaluating the operation of the hydraulics, pneumatics, electrical, environmental and navigational systems. One thing pilots can be proud of, is that they can make sense out of the sea of circuit breakers and instruments that seem to fill every inch of the cockpit. Knowing what this array of dials, CRT instrument displays, buttons and switches do, is enough reason in itself to treat the pilot as a god. Flight planning, paperwork, and for a corporate pilot, catering and baggage handling, all bring the pilot's ego back to the ground. But, once the engines roar to life and the sixty-two thousand pound beast catapults into the air, it's all worth it. We pilots now have total control of not only *our life*, but of *everyone on board.*

Where else can you have total control like that? Passengers can't ask you to pull over and let them out if they don't like your driving. They're in for the duration.

When you meet with some bad weather or an emergency, and your expertise gets them back to terra firma without a scratch, they *finally* realize your worth.

One day I was flying a load of people out of Kennedy Airport to Poughkeepsie, New York. We did this same run four other times that day, but this time we ran into unforecasted precipitation. It was winter and any one knows that cold water freezes when it hits metal. Airplanes are no exception. This airplane could fly with a lot of ice on the wings, but Niagara Falls had just descended upon us without warning. The de-icing system couldn't keep up with the accumulating ice and if we didn't get out of the precipitation fast, we would end up with a quick trip to the trees. We told the passengers we were experiencing a problem and may have to divert to a nearby airport, when we exited the cloud and the ice melted off. Obviously we didn't do anything spectacular. From the passenger point of view though, we were heroes because we solved the problem and kept them alive.

So as I've said before, don't just ask where the bathroom is—give pilots some respect. When you fly, you enter the realm of utter trust to the entities behind closed cockpit doors. How many of you can without reservation give up control of your life? You let someone unseen, take charge of your very being for those airborne hours and you can't say a thing. What is scarier is that some new airliners are flown by computers and as many as *five* computers, which even outwit the pilot sometimes and keep him/her from making mistakes. They really can be foolproof, but now the pilots must be computer literate and VERY computer literate. This is one more reason he/she must be your hero.

✈*Commandment number two:*
Listen.......to the safety announcements.

As boring as this chapter sounds, it is the one to pay the most attention to. Flight attendants, or on newer airplanes, video presentations, bore you with redundant survival information. But—have you ever really listened? Even *one* time? Did you ever read that card in the front pocket? Do you know how to get that oxygen mask on? Do you know how to open all the exit doors, not just the one near you? If you crashed and there was only one way out, you'd better darn sure know how to open that door.

For some information that the safety announcements don't tell you:

✈Count the seats to the exits and put it in your memory.
✈Also count the overhead bins. Lord help you if the airplane crashes upside down and is smoke filled and dark. You will have to use the Braille method to get out and you would be glad you memorized the bin count.

An airplane is an alien environment once it's off the ground. You have to employ your survival techniques, so pay attention no matter who is delivering the passenger safety announcements. When I was a pilot for a commuter airline back in the days where the flight environment wasn't so strict, I had a co-pilot that was only 23 and looked about 15. I wasn't too old myself and was captain of a 19 passenger airplane. On these commuter airplanes, the pilots were required to read the safety information.

One day we were bored with flying back and forth from Kennedy airport to upstate New York and decided to have some fun at the passenger's expense. We kept the cockpit door closed until everybody was on board. I put my hair in two pony tails so I looked even younger. We opened the cockpit doors and made the announcements and then shut the doors. Needless to say, we created 19 sets of white knuckles in the back, wondering if these two "kids" really could fly.

Of course, this form of relieving pilot boredom is frowned upon today, but getting the safety information planted in the passengers' brains is still as important as ever. To illustrate this point I'm going to take you on a little journey. Picture yourself sitting in an airline seat flying along at 35,000 feet. You have just finished your meal and you've decided to delay your urge to run to the bathroom, because the person sitting next to you just opened his laptop. You're thinking about what you're going to do this week end, and you sort of start d r i f i t n g o f f t o t h a t f u z z y s p a c e y o u g e t t o b e f o r e s l e e p s e t s i n.

At least the air is smooth. You think back to some of those flights where it was so bumpy that drinking a cup of coffee was a real adventure. Your head gets heavier and heavier, and you shut your mouth that had unwillingly dropped to its unglamorous open position. The drone of the engine becomes a sleeping pill. It's calm. You're relaxed.

Bang !!!

You open your eyes, feel your adrenaline kick in and your heart is racing a hundred miles an hour! What normally follows is unique to high altitude flying. Simultaneously with the bang, we see all the oxygen masks fall out of the ceiling. This creates ✈ ✈

✈

✈

✈

✈ ✈ ✈

THE GREAT GAPING MOUTH SYNDROME.

The Great Gaping Mouth Syndromes is when:

150 people stare like zombies at little orange pockets that
    are dangling from translucent hoses.
150 people wondering what just happened.
150 people waiting for instructions.
150 people getting woozy.
149 people passing out.

One person listened to the safety announcement and
remembered how to put the oxygen mask on. He's the
person who will survive.

    You only have 30 seconds to don a mask before you
run out of air at 35,000 feet. If you successfully accomplish
this, the dizziness goes away. If you don't, you may become
belligerent or have a false feeling of well being and feel
tingly all over prior to passing out. If you have been
drinking alcohol, your useful consciousness is shortened due
to the depletion of oxygen from your blood because of the
booze.

    If you put your mask on and take a deep breath, you
may feel that you are out of the woods. Unfortunately, this
relative calmness is immediately replaced by your heart
jumping into your throat again as the airplane suddenly
lurches to the right and pitches down! You look out the
window and it seems as if you and one hundred and fifty
other people are going to reach terminal velocity and make
a mighty big hole in the ground. Did you ever hear 150
people scream through an oxygen mask?

    What would you be thinking right now? Insurance
up to date? Will written? Loved ones? Church? God?
Reincarnation? You probably figure death is a definite
possibility don't you? Will you survive? Most likely you
will. Although this is an emergency situation, if handled

properly it is *not* life threatening. The fact that the airplane went into a wing over controlled dive means the pilot didn't die, and is doing what he was trained to do.

When a cabin decompression occurs, the airplane must descend quickly to get you to a breathable altitude. Going straight down is like coming over the top of the roller coaster. Not good for the passengers or the integrity of the wings. The preferred method is to bank the wings, then drop the nose. No negative g's, no wings separating from the fuselage and a whole plane full of thankful survivors. Once you are down to a comfortable unpressurized altitude, you most likely will be making an unscheduled arrival at the nearest suitable airport. Sorry for the inconvenience, but the guys up front just saved your life!

If you weren't afraid of flying before that incident, you probably would be after, unless you knew what was going on. I'm a pilot and I still get nervous when I'm sitting in the back of the airplane and can't see what's going on. Once I was a passenger on a commercial flight back in the old days when they allowed you to hear the pilot to controller transmissions through your audio headset. They also used to have videocams in the cockpit, with the view being shown on the movie screen in the cabin. Well, on this fateful day, I was happily listening to the pilot talk to the controller during his approach to land (backseat flyers could use this as ammunition to criticize the airplane drivers).

Everything seemed normal. We were cleared to land by the tower, and I heard the landing gear extend and lock with a thud. I was sitting in the middle of a widebody so I couldn't look out the window and was depending on the movie screen to let me know when we were close to the runway. It showed us still about 400 feet in the air when I heard another thud. My heart went to my throat. I figured that we had hit a tower or a very tall tree. Then I heard the engines spool up for reverse and my heart slowed down. We were on the ground. I found out later that they commonly froze the camera shot well before the touch down. For pilots riding in the cabin, that was the worst thing they could do for our confidence level. In this case, knowing what was supposed to happen caused me to jump to conclusions and almost have a coronary in the process.

The safety cards in the seat pockets also have lots of useful information. If you are on an over water flight, it may be prudent for you to know where the flotation device is and how to get the life raft out the door. Also depicted is how you should ride the slide during an emergency exit. People get hurt sliding to freedom because they don't follow the rules. It's pretty sad to survive a crash only to be hospitalized because you didn't know how to slide the chute.

These cards also show you how to open the window or the door. I always request to sit by the emergency exit because there is more leg room in this row. The flight attendants will ask you if you are capable of opening the exit window. Don't say yes unless you have read the card. Even if you are not the designated window opener, read the card. You may be the last survivor in a crash and that's no time to figure out how to open the window. Look at these cards every time you fly. Airplanes are different. We all have memory lapses from time to time and a thirty second refresher course in emergency procedures could make the difference in you being a hero or a victim.

✈*Commandment number three:*
Don't complain

Most of you reading this book have probably criticized the person driving when you're sitting in the back of a car. Airplanes are no different. We gripe when it's bumpy. We gripe when we get diverted to an alternate destination. We gripe when we can't take off with the wings covered with ice. We gripe when our flight is canceled for a mechanical. We are very sure we are being inconvenienced purposely and immediately deliver blame where its due—up front.

Criticism number one: *Tossing a few dicey words at the flight attendants because we boarded and then were told we couldn't go.*

If the pilots find anything wrong prior to take off, passengers receive the proverbial announcement. Of course, you deliver criticism at this inconvenience, but isn't it better to know about a malfunction now rather than at 35,000 feet? Sixty-seven percent of the time delays are caused by weather. Frankly, I'd much rather be sitting at the gate when a level four thunderstorm is in progress, than slugging it out in the air with the lightning and turbulence creating a "hell mouth" to fly through. Thunderstorms don't like airplanes and will do anything to spit them out. Years ago a corporate jet tried to out climb a building thunderstorm and lost. He was swallowed up and spit out without wings.

Pilots give weather a lot of respect and will not care a bit about your complaint if they are too scared to go. One fellow pilot relayed a story about a *passenger* who scared *them* into canceling the flight. It seems that this fellow was the only person booked to fly on a commuter flight out of New York. The airplane was a 19 seat Twin Otter. Being designed to fly in remote areas, the cockpit is very utilitarian and even has separate outside doors for the pilots. The passenger (who was the only one on that flight) boarded and waited until the ground personnel shut the door. He then pulled out a gun and tried to hijack the airplane. The pilots told him they had to close the cabin door to hear the radio so the man patiently waited in the cabin—and waited—and waited. Not wanting to be hijacked and not wanting to participate in a game with a gun wielding passenger, the crew just abandoned ship (jumped out the cockpit doors) and left the hijacker sitting in the airplane alone. See, even in this case pilots still have total control.

Once in awhile, after you are comfortably seated, the pilot may tell you that they are having a slight malfunction and are calling in the mechanics. Normally this happens because pilots perform an extensive checklist prior to and during taxi. If something doesn't look right, they delay the flight and call in the mechanics to fix it. If the problem is really bad, you are off-loaded and put on another airplane. This is really annoying, but can you imagine if the pilot said, "the right engine doesn't seem to be developing proper power. Gee, we don't want to disappoint the passengers so I guess we better go and hope it doesn't quit on take-off." Is that what you want? Stop complaining about the delays unless you can clear up the weather or fix the engine.

Criticism number two: *No comforts of home.*

Flying is your time to enter a new world. One that once you are high above the earth, the only reality you have is this tin can with a fixed menu and cramped seats. A memory of life's luxuries exists only in fantasy, unless you can afford to sit in First Class. If you have a problem with fear of tight spaces, you are in for major personal confrontation. How many people have a fear of flying? Believe it or not, many more than ever admit it. If you are employed in a job that takes you on the road frequently, this could be a real problem.

For some reason, when closed into small spaces, people's brains make a decision that they are in peril, whether it's founded or not. Somehow, the body gets the message and creates heart palpitations, sweaty palms, shortness of breath, tight throat, chest pains and a sick stomach. The inside of an airplane is not a fun place to have these symptoms, especially when you are at 35,000 feet and you can't get out. Claustrophobia can become unbearable. Yes, there are things you can take that can alleviate these symptoms and calm your fears and we will discuss them later.

Passengers look forward to the meal service, just to break the boredom. They always gripe about the food and sometimes complain that the flight doesn't stock their favorite brew. The airlines try their best to give you what you want. In the case of beverages, the selections may not be healthy choices for air travel. Since most of the drinks have bubbles in them, they can cause you problems. Since we are flying at high altitudes, air expands (20% or more) which means that the bubbles also expand once they hit your stomach. Now it's called gas and when it wants out of your body, it may not be discrete. One advantage of the air expansion phenomena, is that if you are a male passenger and your seatmate is female with breast implants, her chest will expand about 20% during the flight.

Airline meals bear the brunt of many jokes. They may not be the gourmet dish you prefer, but you must realize that dinners for three hundred people are prepared in advance, shipped to the lockers onboard and heated in galley ovens. Did you ever plan a meal for that many souls and try to serve them all at once from a kitchen that belongs in an efficiency apartment? We complain because they may be loaded with fat, sugar and salt and processed with loads of chemicals, but isn't that the standard American diet? They are just providing us with what they think we want. Your solution to victuals that suit your personal palate, is to brown bag it. Being a vegetarian, I always bring my own chow, since the chefs that supply the airlines, have never seemed to quite grasp what makes a tasty veggie meal. If you truly want the culinary comforts of home, bring it along. Most catering found on corporate aircraft far exceeds the quality found on airliners, primarily because they have fewer mouths to feed and have bigger food budgets. Therefore, if you want an airborne meal fit for a king, climb the corporate ladder.

We also complain that the temperature is too cold, too hot, too stuffy or the cabin air smells. You also may have heard the rumor that flying can give you a cold. Older airlines had an environmental system where air was continuously sucked in and sucked out, giving you a steady stream of relatively fresh air. In most new airliners, cabin air now is recirculated to save energy and fuel. This means the air is not fresh and you can breathe the germs from a sick passenger, over and over. For protection, you can beef up on your antioxidants, wear a personal air purifier and dig into your emergency travel kit explained in the last chapter.

Boredom is another criticism of passengers. Fortunately, the airlines have taken this one to heed. Not only do they supply you with copious copies of magazines, but now you can rent headsets and amuse yourself with the latest tunes, talk radio or a video. You may think you have it bad, but did you ever stop and think what the pilots are doing while you're taking a snooze, working on your laptop or watching the movie? They are NOT taking a snooze or working on their laptop or watching a movie.

Pilots are incredible folks. They can sit idle for long hours, staring into the sun, watching *nothing* move in the cockpit since the autopilot is flying. On nice smooth, uneventful flights their only enroute excitement is when the ground controller's melodious voice issues new instructions. Since not all flights are a piece of cake, the pilots do get to exercise their brain cells when the weather turns sour. Successfully solving an airborne problem and getting you to terra firma in one piece, makes up for all those other boring flights.

Pilots that fly the freighters have a unique way to break the boredom. They catch rats in the cabin. I know first hand that this is true. When you are an airline pilot you can ride for free in the jumpseat of another airliner if there is room. Years ago, I needed to get to California so I hopped on board a 747 freighter in New York. My first inkling that this was not your standard luxury liner was when I met a horse and his trainer who were flying with us. Not to be dismayed at the accommodations, I ran up the stairs to the First Class section.

The cockpit was up there too, and since I was the only free-bee passenger they would let me sit in the cabin where I would be more comfortable. What I found was that this area had only 2 rows of seats and a galley along with three animal traps. I asked the crew what they were for and they told me it was to catch rats.   Rats!!!!!!!!   Rats on an airplane. You bet I didn't sleep on that flight. So, the next time you think of complaining about your airborne accommodations, remember you could be sharing the flight with rats.

For frequent flyers and pilots, the battle of the bulge looms large and many of us complain that the only exercise we get is the run to catch that next plane. Sitting in an airplane seat is not conducive to reducing the fat layers that tend to build up around your mid section, while your fancy ab machine is gathering dust at home. Long flights also manage to create arthritic conditions in even the youngest of passengers. Complaining won't do much for these problems, but you can try hopping up and down the isle, doing squats or deep knee bends as you make frequent trips to the lavatory.

Joining the mile high club at 30,000 feet is not a good exercise option, no matter how exciting it may be. (For anyone who is not familiar with qualifications for membership in that club, it's somehow being able to do the horizontal mambo in an airplane—anywhere in the airplane.) If your hormones get the best of you, think twice before giving in to your desires. Airborne sexual promiscuity is not being taken lightly today. According to the Wall Street Journal, Wednesday June 10, 1998, "Singapore Air claims, that in the previous year, a full one-third of cases of "unruly behavior" requiring cabin crew intervention involved sexual misconduct. In the past two years at London's Heathrow Airport, 15 passengers were hauled in by police for in-flight sexual transgressions." If you are determined to have sex at 35,000 feet, just wait. One international airline is toying with the idea of putting sleeping quarters on transatlantic flights in the future.

Criticism number three: *Why do we have to divert to another airport. What's the matter with the pilot anyway? Doesn't he know how to fly?*

How many times have you been aggravated when you are on a tight schedule and the pilot tells you there will be an airborne delay? This usually comes in the form of a holding pattern. I'm sure most of you have experienced the luxury of extra reading time while the airplane circles around and around. Usually this happens because of bad weather at your destination, or too many airplanes trying to land at the same time.

If the clouds are obstructing the view of the airport, the pilot will make an instrument approach. This means they use radio or satellite navigation to determine their position relative to the runway. By reading the instruments correctly, they will get the airplane right over the runway at the proper time for the wheels to touch down. If the weather is really bad, the pilots may not see the airport when they fly the approach. At this time, they choose to abandon their search for the runway and climb up to a safe altitude. Once back in the holding pattern, they will determine if they want to try again or go to an alternate destination. Instead of being ticked off at the inconvenience you should be glad the pilots made that decision. Try driving a car at 160 mph in dense fog and see how much forward visibility *you* have. The next time you find yourself inconvenienced by an airborne delay, don't complain. Remember, the cockpit crew is using their expertise to get you on the ground in one piece, no matter how long it takes.

Know what all those whistles, whirrs, bumps and clangs are during a normal flight.

The more you know, the less you fear and the better a backseat flyer you'll be. In this chapter we'll go through some airplane scenarios so you can get a taste of what is happening. Your first lesson begins with the engine starting procedure. You'll know this is about to happen when the cool breeze from the air conditioning system stops and the lights flicker. Don't panic. It's just that power is being rerouted for the engine start and all will be back to normal once they fire up. When the airplane leaves the gate and taxis to the runway, lots of things will be moving on the wings. The pilots are testing systems and reconfiguring the airplane with leading edge devices and flaps. These make the wing rounder so it can better support all that weight at slow speeds.

Ever wonder how a 160,000 pound airplane can get off the ground? It seems that whoever created physics deemed it a law that air molecules must always meet up with their friends after passing an object. When the object is

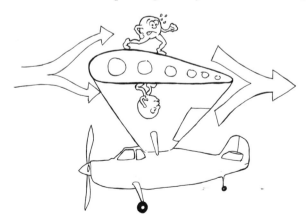

a wing, which is curved on the top side, the air molecules have to *run* to go the extra distance and still catch up to their friends *walking* on the flat under side. The faster the airplane flies, the faster the top guys have to run. These faster moving air molecules create lift on the top of the wing, and up you go.

The air molecules are hardy "folks". They hang onto the wing no matter what, unless the pilot pulls the nose straight up, increasing the pitch angle too much. When that happens, the air molecules lose their footing and fall off. This is called a stall and will cause the airplane to stop flying and fall towards the ground. Once the angle is reduced (the nose pointed level or down), the runners can get a foothold again and you're back flying. Most airplanes out-think the pilots and have buzzers, gongs and a stick pusher to warn the cockpit crew that they are approaching a stall. There are also instruments to tell the pilot if the pitch angle is not correct. Therefore, you really never have to worry about stalling.

Most commercial airplanes have two or more engines. If one engine fails on take off, to keep you safely in the air the pilots perform emergency procedures they learned in the simulator. In their favor is the fact that the other engine(s) is designed to provide enough thrust to keep the wing flying. In small aircraft you may only have one engine and don't want that to fail for obvious reasons, so maintenance and preflight inspections should be absolutely impeccable to find problems before you leave the ground.

I once owned an antique airplane called a Stinson. Because this was a single engine airplane, I was very conscious of every little thing that could make an engine fail. One day I noticed a bird had started to build a nest inside the cowling. I removed the cover and took out what I thought was all of the straw and headed for the wild blue yonder. Shortly after take off we saw smoke coming out of the front of the engine. Fire on an airplane is not a pleasant experience, and all you want to do is get this ship back on the ground. We never saw flames, but when we landed, all the fire trucks rushed out to greet us. It turned out that some of the straw had gotten in between the cylinders and we couldn't see it, so when we fired up the engine the straw started smoldering. Nothing much to worry about, but with only one engine, when you see smoke, it gets your attention.

Little airplanes depend on the pilot to make the lifesaving decisions when a problem arises. Big airplanes have lots and lots of back up systems to help the pilots solve emergencies without the passengers even knowing something went wrong. That's comforting isn't it? These cockpits have annunciator lights, computers and bright bulbs to get the pilots attention when things go wrong. Normally the crew has to do something to fix the problem, although in some new aircraft, the onboard computer acknowledges the problem, fixes it, and then tells the pilot what it did.

Computers are the big thing in airplanes today. Pilots now not only have to know how to take off and land, but they have to be computer literate. Most systems on the airplane are managed by computers including the fuel and the throttle (gas pedal). Even the flight instruments are computer monitor displays. Once the airplane lifts off, the pilot turns on the autopilot. From that point to just before touchdown, the computer flies the airplane. Many airliners are equipped with computers that can even *land* the airplane without the pilots ever seeing the runway until the last few seconds. This is great plus for you when you absolutely have to get to your destination, fog or no fog.

In smaller aircraft, you may not be able to land when the visibility is extremely low because they don't have the expensive on board computer systems of larger airliners. Smaller aircraft depend more on the pilot's expertise. The regulations limit them to certain altitudes at which they *must* see the runway before they land. Don't be upset if they fly right by the airport and go somewhere else where the weather is better. At least they didn't bury you in the trees.

The Take-Off.

Now that you know how the airplane flies and what is controlling it, let's go for a flight. You all are familiar with the take-off. You can't work on your laptop, use your cell phone, radio or other electronic device, must have your tray tables in their upright position and your seat belt fastened. The person next to you may be clutching the armrest tightly with his or her eyes closed. Other people may be reading, giving us the impression that they are apathetic seasoned travelers. This is the moment of truth. You can't get off the ship and whatever fate has in store for you is in the hands of THE PILOTS. You roar down the runway and once the airspeed indicator says you have enough speed for flight, the pilot pulls back on the wheel. Magic happens. This large hunk of metal with 150 trusting souls in it breaks gravity.

When the pilots are sure the airplane won't be landing on the runway again, they retract the gear, which means you hear the wheels go clunk in the wheel wells. Next on the sound agenda are many creaks and groans. These are from the leading edge devices and flaps being retracted. Rounder wings develop more lift at slow speeds so flaps and leading edge devices were invented to change the shape of the wing. When everything is hanging out it's called a *dirty* airplane. When you want to go faster, you need a smoother wing, so the therefore the dirty stuff is retracted and gives you a *clean* airplane. Soon after, you notice the engines spooling back and the noise level being reduced, followed by the plane leveling off and slowing down. Aren't we supposed to be climbing? Is something wrong?

Frequent flyers are used to this even though they may not know why this change occurred. In crowded airspace planes are taking off and landing from the same airport. The ground controller's job is to keep them from colliding. Many times they will hold departing aircraft at a lower altitude for traffic avoidance prior to allowing them to climb higher. At these lower altitudes there are speed limits. If the pilots don't take their foot off the gas when they level off, the airplane will zoom right through the speed restriction and the "Gestapo" will give you a fine. No, there are no police planes with flashing red lights, just the ground controller watching you on radar. If he's in a bad mood he may just report you for speeding.

Once you are clear of the traffic, the controller will give the go ahead to climb. Now the pilot can put the pedal to the metal and you hear the engines get louder again. Other planes are not the only thing you can hit in the air. I once was climbing out of a Virginia airport in our corporate jet. It was a very dark, clear, cold night. This was to be an easy flight home and we were completing our after take-off checklist when I glimpsed a shadow go by the right wing— followed by a bang! We had encountered a kamikaze goose at 7000 feet that did himself in on the leading edge of our wing. He caused $35,000 worth of damage. Airplane windshields are bird-proofed by adding heat which makes the window give upon impact. Wings and tails are not and birds seem to have a big attraction to them. Bird strikes are not normally fatal to airplanes, but they can do enough damage to make you return to your point of departure.

Cruise:

Once you have climbed up to your cruising altitude, the pilot's voice will fill the cabin air, and you are again reassured that a real person is at the helm—or—is it a recording? Things should be pretty smooth for a while. No new noises other than the rustling of the beverage cart and the guy snoring next to you. You can relax for awhile. If you look out the window when the plane turns you can see some neat things happening. Metal panels move on the edge of the wings. They are called ailerons. There's one on each wing and they change the airflow so one wing will lift up while the other falls. This creates a turn. Spoilers and speed brakes can also pop up from the top of the wing. These are designed to also disturb the lift, but in a more drastic fashion. They are used when the pilot wants to descend in a hurry without building up too much speed.

There are two other control surfaces that are attached to the airplane but you can't see them from inside the plane and neither can the pilots. One surface back in the tail is a horizontal stabilizer or elevator. Just as the word describes, it makes the plane go up and down. The other control surface is the rudder which sticks up vertically on the tail and makes the plane's turns coordinated. Did you ever go around a corner too fast in your car? Your body leans to the outside of the turn because of centrifugal force. Do you ever wonder why that doesn't happen in an airplane? It's because the pilot uses the rudder to balance forces in the turn.

Now you have all the knowledge of how the plane goes right, left, up and down. In the cockpit the pilot has a yoke or a joy stick, just like one used with your computer games. In order to control the airplane, the pilots push to go down and pull to go up and turn right to go right and left to go left. That's it. Other than knowing when to speed up or slow down, the only thing left for pilots to learn is how to coordinate all that, land the airplane, fly in the clouds, talk to the ground controllers, solve emergencies, learn navigation and operate the computers. There's an old saying that you could teach a monkey to fly an airplane. Today, if he wanted to fly a modern airplane, he'd have to be computer literate.

The Landing.

Now comes the fun part. The landing phase of the flight. This is where most passengers dole out the most criticism. If you are a pilot sitting in the back, you really have to confront your trust. There is nothing worse than knowing how to fly an airplane and not being able to see out the front window. We backseat pilots press our faces close to the window trying to see around the corner just to make sure the pilot knows where the runway is. If we're flying in the clouds we don't know how close we are to landing until we hear the gear come down. This means that we are about six or seven miles from the airport. If we still can't see the ground we again have to confront our trust level. Pilots have the instruments in front of them so they know where the are, but all you have is hope that they're paying attention.

Before the airplane is in a position to land, it slows down, descends, turns, speeds up, and gets hammered by some turbulence. The flight attendants scurry around picking up the last minute trash and making sure your seat belt is fastened. Over all the commotion the voice from up front gives the proverbial announcement, "put your seat backs and tray tables up." Your first clue that we are close to the airport is the sound of the flaps going down. Remember, we need to make the wing rounder to go slower. It's bad enough that we land this speeding bullet at 120 or 140 knots (nautical miles per hour), but without flaps we may have to increase the airspeed twenty knots or more. Think about that when you know the runway is covered with snow.

At this point in the flight, you should hear the engines getting louder. This occurs because as the flaps go down they cause drag, reducing your airspeed. An airplane can't fly too slow or the little men running on the top of the wing will fall off. Therefore, the pilots have to give her some gas to stay in the air. This also is the case when the landing gear goes down or when turbulence sucks the airplane towards terra firma. All the different ups and downs, acceleration and deceleration may make you nervous, but there is nothing you can do about it anyway. Just sit back and trust that the pilot up front is "on the ball."

Once the landing gear thud is heard, meaning the wheels went down and locked, you are normally less than five minutes from touchdown. If you are sitting by a window you can watch as the ground gets closer, unless of course you are still in the clouds. If the weather is not cooperating the pilot may not see the runway in time to make a safe landing. When this happens, the landing is aborted which means you get to experience the airplane climbing, circling and making another approach. If the weather is really bad, the airplane may go to an alternate airport and you will have to take the bus home.

Pilots rely on radio navigation to guide them down to the field. In many new airliners, the autopilot is approved to fly the whole kit and caboodle to the runway, even if the pilots *never* see the ground. It is during these low-visibility approaches that the airplane drivers up front grow eyeballs the size of baseballs as they try to catch a glimpse of the ground before the wheels do. Now I asked you to trust the pilots. Well, at this point the pilots have to trust the computer. Isn't that a comforting thought? At any rate, most landings are successful. These pilots have been trained up the "gazoo" in simulators to mimic every conceivable unusual situation, and their robotic brains can react in ways that assure you are safely deposited on mother earth.

Once an airplane has contacted the ground it must be stopped. This is no easy task. Jets have reverse thrust which is not, as commonly conceived, where the engines do an about face. Reverse happens because barn doors, called clamshells, close behind the exhaust stack diverting the thrust forward. This effectively pushes the airplane in the opposite direction. Brakes play a part, especially as the airplane slows down. Cars are now equipped with anti-skid so most people are familiar with its design intent. No matter how hard a pilot stomps on the brake pedals, the wheels won't lock up. Good for keeping the you going straight when the runway is slick. Once the airplane stops, you have a unique opportunity to see who was controlling your life. As you exit you may notice the cockpit door will be open. This is your chance to thank the pilot for such a good job. The crew *will* appreciate it!

✈ *Commandment number five:*
Obey the seat belt sign.

When you're up at 35,000 feet, mother nature plays jokes on you. The captain finally turns off the seat belt sign and you make a mad dash for the whiffy (that's pilot talk for bathroom). It seems that just when you are comfortable, mother nature brings in a little turbulence. The seat belt sign goes back on and you're stuck in the can. It usually stays bumpy until you get a real good case of *blue butt* from the stuff in the toilet. Then it calms down long enough for you to make it back to your seat without mishap. The seat belt sign goes off again, but mother nature really has it in for you. She cranks up just enough turbulence to make it impossible to eat, drink, or work on your laptop.

If you don't keep your seat belt on, an unexpected jolt can launch you to the roof and you can damage numerous body parts. According to the National Transportation Safety Board, in the past fifteen years, there have been more than 250 turbulence-related incidents on major airlines. These resulted in three deaths, 70 serious injuries and at least 850 minor injuries. The most publicized has been the recent United Airlines jet upset over the Pacific Ocean that resulted in a fatality. Flight 826, with 374 passengers, fell only 100 feet but caused a g force of 1.8. Lasting only 10 seconds this was enough to uproot anything not tied down, including people! Most airlines are now adopting a policy to keep the seat belt sign on all the time. We all balked at having to wear seat belts in our automobiles when the rule first came out, but now we do it without question. It soon will be accepted in airplanes too. This doesn't mean you are barred from going to the potty, it just lets us know that whenever possible you must protect yourself from turbulence.

Another seat belt phenomenon happens after landing. While you taxi in, "God" speaks and tells you to stay in your chairs. Now, do we listen? No!

We are so eager to get our place in the line that goes nowhere, that we don't even hear the announcement. Wait until some irate captain stops the airplane in mid taxi and says he or she won't move the airplane until you sit down. Who do you think you are to defy the laws of nature? Go ahead, grab that bag out of the overhead bin just when the airplane has to make a quick turn or stop and the guy whose head you drop it on cries law suit. Don't pin it on the airline. They told you to stay seated.

No booze, no caffeine, no carbonation, your own water jug, a pack of gum and a light box.

"No booze? No caffeine? I don't want to read anymore!"

The staples during a flight seem to be booze, caffeine, carbonated drinks, magazines and the movies. Without them, airline travel, as we know it, would be obsolete. BUT—if you care about your health and if you care about your looks, you may want to keep your eyeballs pasted to this chapter.

When the beverage cart travels that long lonely isle dispensing its boredom chasing liquid refreshment, you have a choice. Remember, airlines are there to satisfy you, and you want booze, caffeine and soda. Let us elaborate and maybe you will ask for juice or water instead. Booze poisons the tissue cells. This prevents our bodies from utilizing oxygen, which is hard enough to come by since most cabin altitudes are about as high as the base lodge of a Rocky Mountain ski area. If you're like me, the first day you arrive in the mountains you sleep, and sleep, and sleep. That's because of the thin air (a lack of copious amounts of oxygen). In airplanes you have the same problem of not getting enough oxygen, so why would you want to compound the problem by adding alcohol? By the way, if you want a cheap drunk, imbibe at altitude. Alcohol's impact on the body is two to three times more potent when you're flying. One glass of wine aloft is like three on the ground!

The problem with caffeine is that it dehydrates you. Since a high altitude airplane is almost devoid of humidity, the only moisture you get is from other people's sweat. As the moisture evaporates from our bodies our skin dries out. Did you ever look at an airline captain who has spent twenty or thirty years in the cockpit? Is he gray and wizened with a face that looks like a well used leather bag? Many are, and most of them didn't pay attention to the need for rehydration at altitude. Because of his seasoned appearance, you may feel more confident in his ability to pilot your plane. But, just think of what *you* will look like if you're a frequent flyer that substitutes caffeine drinks for good old water.

'Ah, that's OK. I'll just have a seltzer with lime.' Wrong! That falls under the category of carbonated drinks, along with soda (or if you're from the south, pop). As we explained earlier, air expands 20% or more up at altitude and that includes the bubbles in your drink. So what, you say. I can handle the gas in my stomach. I have my little roll of antacids. What they don't tell you is that those bubbles can play havoc on your digestion. The phosphoric acid also has the unique ability to leach calcium out of your bones. Baby Boomers beware! Frequent bubbly drinks can help you on your way to osteoporosis.

So now that I have taken away most of the fun you can have on the airplane, you are faced with the last choice. Juice and water. Sure you can put that lime in the water and pretend it's a tasty drink. Maybe you'll find out it *is* a tasty drink, and a lot better for you. Don't tell me you aren't able to go four or five hours and just drink water. This brings us to the water jug. Airliner water comes from the hose at the terminal. It is probably polluted and chlorinated (which is very bad for you). Plus—who knows when the last time was that they disinfected the water tanks on the airplane. Fortunately, the airlines have your health in mind and can offer you bottle water. This is your best choice. You can also elect to carry your own bottle of *purified* water. This is a trendy thing, but on long flights you may run out. Remember, you must drink *lots* of water to keep those wrinkles away. Many airlines provide bottled water, and some may offer you a new product that not only gives you purified water, but it's loaded with electrolytes and antioxidants. This is important as it keeps you hydrated while helping you fight onboard germs.

Now the good part. You can have gum. The gum is not a replacement for a meal, nor to feed your sugar fix, but to help you clear your ears during the descent. It also breaks the boredom and helps smokers make it to the next stop. If you don't like gum there are alternatives for clearing your ears. You can hold your nose, close your mouth and attempt to blow out to neutralize the pressure. There are ear pads that you fill with warm water that may help. Decongestants are not a good choice as they dehydrate the sinuses that are already dry from the lack of cabin moisture. Herbal nasal sprays are better choices as they reduce the internal swelling without dehydration. Flying with a cold or severe allergies is a no, no. When our sinuses are stuffed up our ears cannot adjust air pressure. The result is torture. I have experienced the most excruciating ear pain in my life during the descent. I now understand why all the babies on the flight start screaming as we descend for the airport.

For those who only fly north and south, you can skip to the end of this book. For the rest of us, this section can alleviate one of the most exasperating side effects of flying, jet lag. Diana Fairechild wrote a great book called "Jet Smart." She meticulously schools us in combating jet lag. Because of her labor I won't have to delve into this subject very much as you can just read her book. There are a few pointers however, that we would like to give you regarding this subject.

Most travelers flying east experience jet lag more often that those flying westbound. Jet lag is a biological disruption caused by flying across time zones faster than the body can adjust. Our biological (circadian) rhythms are messed up and our internal body clock becomes confused when it thinks the sun is supposed to be out and it's dark instead. When this occurs, the body compensates in some very undesirable ways

Many people don't know the symptoms of jet lag, except that you seem to either want to sleep all the time or experience annoying insomnia. Some less well known symptoms are:

✈ anxiety
✈ body aches
✈ broken sleep patterns
✈ dehydration
✈ depression
✈ dry cough
✈ disorientation
✈ fatigue
✈ headache
✈ impaired vision and coordination
✈ impatience
✈ inability to concentrate
✈ indecisiveness
✈ irrational behavior
✈ lightheadedness
✈ loss of sexual impetus
✈ memory loss
✈ nausea
✈ swollen feet

These are not really symptoms you want to bring to that high level board meeting. Symptoms of jet lag are relatively easy to address with a few key single homeopathic remedies including Arsenicum album, Crataegus and Ginseng. The latest and trendiest fix for jet lag is to carry your own personal light box. To quickly effect a remedy you must reset your circadian rhythms and reset our body clocks. The most important regulator of these is the daily alternation of light and darkness. In plain English, our bodies are controlled by our internal clocks that base everything on sunrise and sunset.

NASA uses carefully timed bright light exposure to reset the sleep-wake cycles of astronauts. This doesn't mean that you look directly at the sun or stare at a light bulb. It means being exposed to a light source 20 times brighter than normal indoor lighting at specific times. Proper usage of a light box can reset your body clock within one day. Normally to balance your circadian rhythm, it would take up to three weeks. This new technology can also reduce the SAD Syndrome (seasonal affective disorder). SAD occurs during the late fall in the northern hemispheres (late spring in the southern ones), when the light vanishes at 4:00 p.m.

Roger J. Cole, Ph.D., conducted one of the first scientifically controlled trials of bright light treatment of jet lag. From 1987 to 1992 he studied the effects of light on human biological rhythms, at the University of California, San Diego. He has developed a light exposure calculator that easily allows the traveler to determine when to use the light box. For example, if you crossed three time zones going eastbound, you should avoid light before 8:00 a.m. and seek light after 8:00 a.m. On westbound flights over three time zones, you would seek light before 2:00 a.m. and avoid light after 2:00 a.m. Since it's hard to find light at midnight in most latitudes, you need a light box. Through nerve pathways from the eyes, the light from the box at 10,000 Lux, rapidly modifies the internal clock located in an area of the brain called the suprachiasmatic nuclei (SCN).

A word of caution, if you don't follow the light exposure timing properly you may reset your internal clock in a direction opposite to the one desired.

Many of these light boxes are so compact you can carry them on the airplane. For corporate travelers, the company may just so be inclined to provide you with one when you reach your destination, in order for you to quickly get back in top notch "business" form. In the future airlines may install bright lights on their planes to help travelers adjust to the new time zone. This is just part of the airline's desire to nurture the passengers and keep them coming back.

Along that same line, instead of looking at the light, could we wear light boxes around our knees and get the same results? In an experiment, scientists have shone a bright light on the backs of human knees and reset their biological clocks. Dr. Michael Menaker, a biologist at the University of Virginia in Charlottesville was quoted in the New York Times on January 16, 1998. "For three days we tried to find flaws in the experiment and we couldn't." The study mentions that other creatures have light sensitive cells on their bodies; shoe crabs have them on their tails, swallows inside their skulls and fruit flies have sensors all over their bodies. The study shows that we may not be limited to our eyes for light sensitivity. Wearable light boxes may someday be the hottest item in the airport gift shop.

Speaking of things you wish were in the airport gift shop, how about a pill for fear. Our next chapter specifically deals with health issues and fear of flying can cause some pretty nasty symptoms.

SERIOUS STUFF:
YOUR SURVIVAL TRAVEL KIT

Many of us never seem to feel the effect of air travel, or at least we don't *think* we feel the effects. Maybe those aches and pains and illness we seem to get as we age, started on those frequent flyer trips. This chapter may open your eyes to things that can protect you in the alien environment of 35,000 feet, including solutions for air sickness, catching cold and the fear of flying.

There are numerous fearful flyers and even more numerous members of the fear of flying club that don't admit it. Many passengers have difficulty letting someone else (the pilot) control their life. When times get tough and they have to trust someone other than themselves, they fall apart. No matter what anyone says, the symptoms you get when you become anxious while flying, are *real*. Some reactions to aerophobia are sweaty palms, sweaty bodies, heart palpitations, claustrophobia, difficulty breathing and a relentless desire to get out of the airplane. We found several videos available on the internet that deal with conquering the fear of flying. They explain how to effect long term cures and are well worth the money. In this chapter, we let you know how to get *instant* relief from the anxiety.

The average person doesn't understand what you go through when you experience this type of panic. Just when you think you've conquered the flight, the airplane taxis in and stops, but the door doesn't open. 150 fellow passengers are packed in the isles complaining that they will miss their connections. You're safely on the ground so what are you worried about? Claustrophobia hits and you picture yourself clambering over bodies to be first in line to get out of this sardine can. You start sweating, your heart starts pounding and you feel like the oxygen has just been sucked out of the cabin. You really think you are about to die. These are real symptoms and no matter how much we try to think them away, they persist.

In 1980, The Boeing Company published a report by Robert D. Dean and Kerry M. Whitaker entitled "Fear of Flying, Impact on the U.S. Air Travel Industry." The report indicates that one of every three adult Americans is either anxious or afraid to fly. Of those afraid, 73% were frightened of in-flight mechanical difficulties, 62% of bad weather, 36% of on-ground mechanical problems, 33% of over water operations and 36% of night flying. They didn't include female baby-boomers who were going through menopause and experienced panic attacks for none of the above reasons. I was one of those women. Since flying was my livelihood and I definitely was *not* afraid of it, these symptoms confounded me. I delved into research and came up with some solutions.

Since many panic attacks can be caused by a release of adrenaline, sometimes caused by low blood sugar, it is important to eat before a flight. The meal should be of protein (meat or vegetable) and carbohydrates. No sugar should be taken before or during the flight as sugar spikes adrenaline responses. This includes soda and alcohol which provide sugar highs. Caffeine also can provide a response similar to an adrenaline rush. Therefore, colas or coffee are not good beverage choices. On long flights it is prudent to eat every three hours. If the airline doesn't serve meals that frequently, brown bag it, but don't munch on candy bars (they contain sugar).

For those of you familiar with homeopathic reme-dies, try gelsemium as a relaxant. There also are herbal tinctures you can carry with you that reduce anxiety; Elixir of Passionflower which acts as a sedative; Skullcap, normally compounded with St. Johns Wort to act as a relaxant; Lemongrass and Kava Kava, that calms down the central nervous system; Valerian Root, which acts as a muscle relaxant. If you have used Valerian before and liked it, you may also think of trying an herbal combination of Valerian and Wild Lettuce. These two herbs complement each other by providing a very strong relaxant for extreme cases of panic. It directly balances the release of adrenaline by providing a sleepy feeling of complete relaxation.

Accupressure can be used to alleviate the panic and you can do it on yourself. One point is on the crease of the inner wrist, in line with the smallest finger. Another is slightly below the level of the bottom of the kneecap, on the outer side of the leg in the slight depression formed when the knee is completely straight. The technique to use is to rub with gentle finger pressure. The spot may be sore, but as you rub the soreness and the anxiety should diminish.

My favorite remedy is extracted from flowers. It works in seconds, right in the middle of a panic attack and stops it cold. The combination flower remedy tincture I use is called Fearfulness™. It contains Aspen, Blackberry, Cherry Plum, Garlic, Mimulus, Red Chestnut and Rock Rose. The remedy addresses the fear of flying by directly bringing up and calming the underlying emotional issues that cause the fear whether you know what they are or not. On the package it says this remedy is for fear of darkness, being attacked, aging and dying, being alone, going crazy or loss of control, dreading life or feeling panicky and terrified. However, if you aren't afraid, but just apprehensive, you can try another flower remedy combination that combines Agrimony, Aspen, Bottlebrush, Impatiens, Red and White Chestnut and Chamomile. Since panic attacks begin in our emotional thought patterns and our mind, it seems logical that we should treat the cause, not just the symptoms.

Speaking of symptoms, have you ever used the barf bag? It's not pleasant for either you or the person sitting next to you. Air sickness is as miserable as sea sickness. Yes, people have been known to actually turn green during their bout with the equilibrium malady. Air sickness is caused by the inability of the mind to make sense out of what the eye sees and what the ear feels. You are flying in the clouds in a tube where the structure looks like it should be vertical. Instead your ears tell you that your body is leaning at 30 degree angle. In the middle of this confusion is your stomach, which doesn't like the discrepancy and prefers to get rid of its contents in rebellion.

There are certain drugs that can prevent airsickness but we prefer alternatives. A homeopathic remedy for vertigo (ear/eye inconsistencies) is Tabacum. Ginger has been used as an old home remedy for sea sickness and car sickness and is perfect for air sickness. Ginger tea or ginger powder on the tongue is best, but ginger cookies will do in a pinch. Other effective herbs in combination are Ginkgo Biloba Leaf, Passionflower, Clove Buds, St. Johns Wort, Lavender Flower and Hyssop Herb to name a few. These remedies should be taken for five days prior to travel to be most effective. If you become nauseous on the plane, see if someone has peppermint. Oil of peppermint without the sugar is best, but not many people carry that aboard, so search out peppermint gum or candies.

A great help to stomach upset is to take along some digestive plant enzymes. As we will describe later in this chapter, these should be part of your immune system travel kit, but they also have the ability to settle your stomach. There are herbal combinations that supply nutrients to help balance the body's equilibrium and reduce trauma in the stomach. Herbal formulas that contain liquid extracts from Ginger, Cassia, Clove Bud, Hyssop, Black Caraway Seed,

Nutmeg, Peppermint and Red Poppy flower can do the trick.

A gizmo that helps you avoid air sickness and sea sickness has found acceptance in mainstream circles. This is a cloth band that you wrap around your wrist that has a little ball embedded in the band. You place this so the ball ends up three fingers up your arm from the crease on your wrist. This is an old accupressure trick and is most widely used now to prevent sea and air sickness. It is extremely effective.

If you are one of the lucky travelers that doesn't panic and doesn't get queasy, you still may be a candidate for catching a cold. Airplanes maintain their environmental systems by taking air from the engines and routing it into the cabin. This air often is recirculated through the cabin several times before it exits the airplane and fresh air replaces it. What this means to the captive passengers, is that if someone sneezes, those germs may hang around for quite some time before being expelled into the atmosphere. If you have an immune system that has temporarily suspended operation, you are a viable home for those germs. Most of us have compromised immune systems due to the type of environment we live in, and the type of diet we choose. During research for our many health books, Dr. Howard Peiper and I discovered three key supplements that give our body what it needs to successfully fight illness. These three need to be part of your travel kit.

The foundation of the immune system is *minerals.* Without sufficient minerals, the body cannot operate properly and our immune system goes into hibernation. Both our water and the soil our food is grown in are deficient in minerals. Therefore it is absolutely necessary to take a mineral supplement. I personally use a liquid crystalloid electrolyte supplement that I add to the water bottle I carry with me. If you want to know the nitty gritty about minerals, read our booklet, "Crystalloid Electrolytes," listed at the end of this book. A good shot of minerals at the first sign of a cold, may just wipe out the germ.

The next most important supplement to carry with you is *digestive enzymes*. Only raw foods contain digestive enzymes and since most of us don't eat a totally raw diet, our bodies need help in breaking down our dinners. Normally the pancreas does this job, but the more cooked food we hand it the more it becomes tired and lies down on the job. This results in food not being digested and nutrients not being absorbed by the body. Your immune system suffers because it's not absorbing nutrition from the food. Now you are more susceptible to every cold that comes around the corner or to the sneeze in the next airplane seat. Plant enzymes are the best to carry as a supplement because they work throughout the whole digestive tract. You need to add enzymes to your diet every time you eat something cooked or processed, even snack foods too.

The third item that you need to support the immune system is an *essential fatty acid supplement*. Your brain needs this good type of fat in order to function properly and tells the immune system how to work. Especially if you are on a low fat diet, you may be deficient in the good kind of fats. Taking along a bag of ground flax seed to sprinkle on cereal or salad is a simple way to make sure your brain has its food. Flax is available in capsules if that is more convenient for you.

I can't stress too much the importance of supporting the immune system. By taking only these three supplements you have an excellent start on protecting yourself from the germs circulating around inside an airplane. If you really want to kill that cold germ once you suspect that it has infected you, try an herbal antibiotic. Echinacea with

Goldenseal Root comes in a tincture or a tea and is an excellent way to get rid of a cold fast without using drugs. As a safety measure, add it to your travel kit. You can also help to detoxify the harmful pathogens from your body by stimulating the proper function of the liver. Three of the best homeopathic remedies for this purpose are Bryonia, Chelidonium and Nux vomica.

Beyond taking supplements, there is a high tech device that you can wear around your neck that will keep the germs away from you. This device employs three technologies, electron impact decomposition, magnetic field enhancement and corona discharge ionization. It doesn't matter that you don't know what any of that means, just remember that it can clean the air around you and keep those germs from invading your space.

The ionization and magnetic field have the added benefit of creating a calming atmosphere that can help reduce anxiety and jet lag. This premise is based on a hypothesis by Richard H. Lee, P.E., in his book "Bioelectric Vitality." He states, "an airplane, because of its position above the earth and its metal structure, decreases the availability of both electrical and magnetic substance from that which we experience on the planet's surface." As magnetic energy in the passenger decreases, energy becomes unbalanced resulting in sensations such as anxiety, nausea, sweating, rising heat, itchy eyes and headaches.

Fewer electrons, or free negative ions, are available which contribute to the emotional lack of well being. No wonder fearful flyers get more fearful once they are up in the air. Lee also uses a magnet that hangs down the back of your neck, with its negative pole facing the base of the neck at C-7 (cervical location on the spine). He finds that this decreases anxiety and removes the tension that builds up in the back during airplane flights, a common complaint of frequent flyers.

Another "Star Trek" device that you can own protects you from the harmful effects of electromagnetic fields (EMF's). It is well known that EMF's coming from a computer can make the operator extremely fatigued and suppress the immune system. They have also been implicated in causing miscarriages and even cancer. Airplanes are full of EMF's because of all the electrical wires running through the fuselage, and this can be the primary reason you get tired on an airplane.

These electromagnetic fields are chaotic and when they interact with your body's electricity, a "short circuit" occurs. Your body continually fights to reestablish a harmonious frequency. Repeated assaults cause your body to work overtime and it eventually gets pooped. All this, while you are doing nothing but sitting. To solve the problem of EMF's, you can take along your elemental diode.

This little device handily fits in your pocket or can be purchased embedded in a bracelet. It emits forty-seven resonant frequencies that synchronize and correct the energies flowing into the human body. This helps balance the body's electrical system. I personally have used this and can attest to its ability to reduce fatigue.

A common side effect of traveling is irregular bowel movements. We never talk about it, but most of us have trouble with constipation when we do our business in unfamiliar surroundings. Much of this trouble comes from our erratic eating habits and having to substitute restaurant food for home cooking. One of our favorite solutions is to take along a high fiber, chlorophyll cereal grass powder. When mixed with your morning juice, it can become your answer to this uncomfortable side effect of traveling.

Listed below are the basics we encourage you to use if you are a frequent flyer. *Starred items are a must for all travelers that want maximum protection regardless of personal condition.* As we've said before, the inside of an airplane is an alien environment and you must protect yourself. If you neglect to heed our recommendations, we will not be responsible for the ailments you encounter from air travel, nor the wrinkled face that could mark you as a frequent flyer. Flying should be a pleasant experience, not a health hazard. After all, how many times do you get a chauffeur to drive you 3000 miles, serve you a meal, offer you a movie and let you snooze, without asking for a tip.

Air Travel Survival Kit:

★Mineralized water
★Digestive Enzymes
★Flax
  Light box
  Natural supplement for jet lag symptoms
  Natural air sickness remedy
  Protein snack (no sugar added)
  Herbal antibiotic
★Personal air purifier/ionizer
★Elemental diode
  Natural nasal spray
  Wheat grass powder
  Natural detoxifier

EPILOGUE.

Although this book was meant to be a light dissertation on flying as an adventure, we did throw in some serious stuff. The inside of an airplane is no less friendly than the inside of a submarine. You cannot exist outside of either without special protection and if the outside suddenly comes inside your ship, you'd better know what to do. Even if the flight is uneventful, you still can benefit from the information you received in the last few chapters.

The best part is that with the knowledge you gained from reading this book, you are now considered qualified to proudly wear the honorary title of

"BACKSEAT FLYER"

Congratulations!!!!

RESOURCE DIRECTORY

Books that give you information about solving a particular health problem, but don't tell you where to get the stuff, really frustrate me. This is not that kind of book. Our Resource Directory chapter will give you some companies to check out if you are looking for products that solve some of the problems associated with flying, such as jet lag, airsickness, fear and staying healthy.

EARN TRAVEL MILES ON TOP OF MILES. Join the International Airline Passengers Association (IAPA) and earn frequent flyer miles accepted by 18 different airlines! Save up to 50% at over 5000 hotels worldwide and 30% at virtually all major car rental agencies. Protect your missing bags with the unique Bag-Guard® retrieval system. Free reservation service, exclusive members-only newsletter and more! IAPA, P.O. Box 700188, Dallas TX 75370 (800) 821-4272

STOP EAR PAIN WHEN YOU FLY. Anyone who has flown with a stuffy nose is aware of the problem of plugged ear canals during the descent. Formulated to help ear problems is *Ear-Lyte*™, an all natural herbal and electrolyte blend of drops to give rapid relief to ear problems. To avoid this pain, without the drying effect of decongestants, we also suggest herbal electrolyte *Nasal-Lyte*™. Designed from an old Naturopathic remedy, it consists of a blend of 10 medicinal herbs that soothe and penetrate the deepest recesses of your sinus passages, virtually offering instant relief. NATURE'S PATH, INC., P.O. Box 7862, Venice FL 34287-7862 (800)-326-5772

ENZYMES FOR IMMUNE SUPPORT. The lack of enzymes in our cooked-food diets hamper proper digestion. This limits the nutrient absorption needed by our bodies to support our immune system. *TYME ZYME*™, an all natural scientifically proven formula contains all the necessary enzymes for digestion throughout the intestinal tract. It contains protease, amylase, lipase, cellulase and lactase. When taken with meals, it increases nutrient absorption and assures the body of receiving the benefits of vital nutrients and essential fatty acids. This strengthens the immune system, aids in digestion and increases energy. PROZYME PRODUCTS, LTD., (800) 522-5537 call Debra Casey for information.

BRAIN FOOD FOR YOUR IMMUNE SYSTEM. As necessary ingredients for proper cellular neurotransmitter function in the brain and throughout the body, Omega-3 essential fatty acids must be balanced with Omega-6 essential fatty acids. Flax provides a good balance of these nutrients. *Fortified Flax* contains Organic Flax seed, Zinc, Vitamin B-6, C, E and is "yeast free". For a healthy snack, they also offer flax in a tasty *Omega Bar*, a convenient way to get your energy. *Fortified Flax* and *Power Pack Energy Drink* Mix can be sprinkled on cereal and sandwiches or mixed with juice or water. OMEGA-LIFE, INC., P.O. Box 208, Brookfield, WI 53008-0208 (800) EAT-FLAX (328-3529)

HOMEOPATHIC REMEDIES. To relieve symptoms of jet lag the combination homeopathic *#24 EXHAUSTION* is effective. Take three to six drops or pellets on the tongue one to three times a day. Use in combination with #1 *DETOXIFIER* complex for long standing fatigue. When confronted with airborne pathogens such as those found

trapped inside an airplane, the #1 *DETOXIFIER* is used to support the liver in preventing common ailments. Daily suggested use is three to six drops or pellets on the tongue at bedtime. NEWTON LABORATORIES, INC., 2360 Rockaway Industrial Blvd. N.W., Conyers, GA 30012 (800) 448-7256 Email: mailinfo@newtonlabs.net
Website: www.newtonlabs.net

DON'T SUFFER FROM JET LAG. Since 1986 Apollo has helped NASA astronauts, pilots and travelers reset their body clock and eliminate the effects of jet lag through use of bright, artificial light. The *Travel Lite* is 20 times more powerful than indoor light, weighs only 3 pounds, is 100% safe, UL listed and tricks the brain into resetting your body's internal clock. The *Jet Lag Calculator* helps you compute which time of the day you need to receive or avoid bright light. (Free with Travel Lite purchase.) 30% discount available by mentioning "The Backseat Flyer." APOLLO LIGHT SYSTEMS, INC., 325 West 1060 South, Orem, UT 84058 (800) 545-9667

FRESH FROM THE FARM. FLAX FOR YOUR IMMUNE SYSTEM. A whole food, *Dakota Flax Gold* is all natural edible fresh flax seed, is high in lignins which can be used over cereal, on salads, in soups or in juice. Ready to grind, just like your best coffee, it is low in cadmium and is better tasting than packaged flax products. Seeds must be ground for full nutritional value. Dakota Flax Gold is available with grinder. Flax, also available in capsule form as *Flaxeon Jet,* is a convenient way of getting beneficial essential fatty acids. HEINTZMAN FARMS, RR2 Box 265, Onaka SD 57466 (800) 333-5813 (send S.A.S.E. for sample)
Website: http://www.heintzmanfarms.com

FLOWERS & HERBS FOR AIR TRAVELERS. *Fearfulness*, a homeopathic-style Deva Flower Remedy® effectively addresses fears of flying by directly bringing up and calming the emotional/mental issues that cause the fear. *Fearfulness* will help calm those feelings, aiding in the release of underlying motives behind the fear. Equally effective, *Motion-Stop* a Medicine Wheel Herbal Drop™, is used to offset jet-lag and air sickness by balancing the body's equilibrium, thus calming the body and stomach. It's effectiveness is obvious when taken for a period of 5 days prior to travel. NATURAL LABS CORP., P.O. Box 20037, Sedona, AZ 86341 (800) 233-0810 Email: Natlabs@sedona.net

OVERCOME ANXIETY, FATIGUE AND JET LAG. Thousands who used to be terrified of air travel use the *Air Travel Vitalizer*. Not only does this small device kill airborne viruses and bacteria to protect you from illness, but it also fills your lungs with magnetic energy to reduce anxiety, and dizziness. Worn around the neck, the *Air Travel Vitalizer* can reduce the symptoms of jet lag and those associated with the fear of flying by emitting fresh, ionized air. It is also excellent for anyone who has asthma or environmental sensitivity, and can be used at home, work or in crowds. CHINA HEALTHWAYS INSTITUTE, 115 North El Camino Real, San Clemente, CA 92672 (800) 743-5608

AIRBORNE ELECTROLYTE WATER. There is no substitute for water as the health drink of choice while flying. When that water contains electrolytes (the spark that charges your body's battery), and antioxidants, there is no better drink to keep you healthy and better able to avoid jet lag and airborne colds. *Flight-Lyte*, a bubble-less brew contains both crystalloid electrolytes and antioxidants. It is available in easy to carry serving size containers and even may be available on the beverage cart of your favorite airliner. OCEAN-LYTE ENTERPRISES. P.O. Box 531, Jenison, MI 49429-0531 (888)-NUTRITION

HERBAL REMEDIES FOR AIRBORNE MALADIES.
Traveler's Comfort is an herbal combination formulated to help with motion sickness, One of its ingredients, Ginger Root helps ease dizziness and nausea by reducing the sensitivity of the motion-detecting nerves in the ear. *Ginkgo Biloba* is another herb proven to help balance the inner ear and increase circulation to vital organs that are placed under stress during motion. *Relax Caps* is a combination herbal formula with a potent blend of Kava Kava and Valerian root designed to ease the anxiety and stress associated with fear of flying. CRYSTAL STAR HERBAL NUTRITION, 4069 Wedgeway Court, Earth City, MO 63045 (800) 736-6015

HELP FOR JET LAG, FATIGUE AND HEADACHES.
Research indicates that these and other symptoms may be caused by electromagnetic fields (EMF's) emitted from electrical equipment and the wiring in airplanes. The all natural elemental Diode formula successfully neutralizes these EMF frequencies that are harmful to the body. Scientifically tested by the FDA-registered Phazx and Voll Dermatron Machines, the Diode formula frequencies have produced excellent results. The diode can be worn on the body or placed on devices that emit electromagnetic radiation. ENER-G-POLARI-T PRODUCTS, P.O. Box 2449, Prescott, AZ 86302-2449 (800) 593-6374

Bibliography

Anderson, Nina, Peiper, Howard, <u>Over 50 Looking 30</u>, Safe Goods, E. Canaan, CT, 1996

Blakeslee, Sandra, <u>Study Offers Surprise on Working of Body's Clock</u>, The New York Times National, Friday, Jan. 16, 1998

Brown, George Albert, <u>The Airline pasenger's Guerrilla Handbook</u>, The Blakes Publishing Group, Washington, DC, 1989

Business & Commercial Aviation, <u>Intelligence</u>, May, 1998

Cole, Roger J., Ph.D., <u>New Light on Jet Lag, The Fast Way to Reset Your Body Clock,</u> Light Exposure Calculatro, Orem, UT, 1995

Detwiler, Ross C., <u>Dichotomy</u>, Business & Commercial Aviation, August 1995

Fairechild, Diana, <u>Jet Smart</u>, Celestial Arts Publishing, Berkeley, CA., 1992

Lee, Richard H., P.E., <u>Bioelectric Vitality,</u> China Healthways Institute, San Clemente,CA, 1997

Lewy, Alfred J., MD, PhD, <u>Treating Chronobiological Sleep and Mood disorders with Bright Light</u>, Psychiatric Annals 17:10, October 1987, p664-669

Nomani, Asra Q., <u>Plane Misbehavior,</u> The Wall Street Journal, Wednesday, June 10, 1998, pg 1

Rosenthal, Elisabeth, <u>Light Pulses Shift Astronauts Rhythms,</u> The New York Times Science, Tuesday, April 23, 1991

INDEX

OTHER BOOKS AVAILABLE FROM SAFE GOODS

OVER 50 LOOKING 30! The Secrets of Staying Young. $ 9.95
How to become wrinkle resistant and fight the signs of aging.
All Natural Anti-Aging Skin Care $ 4.95
The newest information on keeping your skin young.
The All Natural Anti-Aging Diet $ 4.95
Eat lots, Stay slim and avoid old age diseases.
A.D.D. The Natural Approach $ 4.95
Alternatives to drug therapy for children and adults with attention
  deficit disorder.
The A.D.D. and A.D.H.D. DIET! $ 9.95
 look at contributing factors, natural treatments for ADD/ADHD.
Put Hemorrhoids and Constipation Behind You $14.95
A natural healing guide for easy, quick and lasting relief.
The Humorous Herbalist $14.95
Practical guide to leaves, flowers, roots, bark and other neat stuff.
The Brain Train $ 4.95
How to keep our brain healthy and wise (for children).
Natural Solutions to Sexual Dysfunction $ 9.95
Natural alternatives to drug therapy for sexual problems.
Growth Hormone, The Methuselah Factor $12.95
Reverse human aging naturally.
A Guide To A Naturally Healthy Bird $ 8.95
Nutritional information for parrots and other caged birds.
Super Nutrition for Animals (Birds Too!) $12.95
Healthy advice for Dogs, Cats, Ferrets, Horses and Birds.
Plain English Guide to your PC $ 8.95
The computer book that tells it better and tells it in English.

Order Line (800)-903-3837
Safe Goods Publishing
PO Box 36, E. Canaan CT 06024
860-824-5301
www.animaltails.com